MW01493030

TIMELESS
TRIOPS

A PREHISTORIC CREATURE

A Science + Art Book by Lori Adams

Timeless Triops
A Prehistoric Creature
ISBN-13: 978-0-9895368-0-6
Library of Congress Control Number: 2013917424
© 2014 Lori Adams Photo
Published by Lori Adams Photo
Hopewell Junction, New York
www.loriadamsphoto.com

Printed in the United States.

Dedicated to the first artists and teachers in my life: Eloise Adams, an English Major who still loves words and stories, and Jack Adams, who knows about cameras and industrial arts and was and is still patient with questions as to what things are and how they work.

TIMELESS TRIOPS

A PREHISTORIC CREATURE

Message to Readers

Dictionaries have always been a terrific resource in my life. And as I learn about biology and zoology, my vocabulary grows in leaps and bounds! There are so many words and terms ("verbal shortcuts") that describe complicated ideas. These words are like successful photographs, illustrations or diagrams, because they invoke volumes of information.

May you enjoy the many wonderful words introduced in *Timeless Triops* as well as the illustrative images. I hope dictionaries are or become a lifelong source of satisfaction and joy for everyone who is reading this book now, or hearing it read aloud. I also hope you become as enthralled and enamored as I am by these tiny treasures called Triops.

Lori Adams

What is a Triops? What are Triops?

If you don't know the answer to this question, you are not alone. A Triops is a creature that is not found in backyards, zoos, lakes, rivers, sewers, forests, farms, suburbs or cities. Triops are freshwater crustaceans. They are mainly found in deserts and then only during rainy seasons.

Does a Triop exist?

Well, not really. One Triops is still a Triops. The name of this animal has an *s* on the end of the word, always. It doesn't matter whether there is one or two or a million of this creature—it or they are always called Triops.

What does it mean to be timeless?

Basically, it means without time or not having a beginning or end. It can mean "forever" or "ageless." It can also mean that it doesn't change, that it lasts or endures and keeps going and going and going.

How big is a Triops?

The kind of Triops pictured here, *Triops longicaudatus* (meaning "longtail"), grows to about 2½ inches long, including the tail portion.

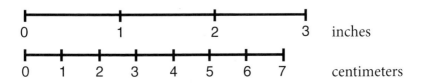

What does "prehistoric" mean?

When the word *prehistoric* is separated into parts, you get "pre" and "history." "Pre" means before and "history" can mean what has happened before now or what is in the past. This is a strange idea! Before before? So in a very special way, this concept really fits this unusual animal.

The word *history* can also mean the telling of a story, or that written records (in words or pictures) have been kept about a particular subject by human beings. In this case, prehistoric means a time before now, when there were no written records kept of anything by humans.

Where did Triops live? Where do they live now?

Triops have always lived in deserts during the rainy season or in pools or small ponds of water that come and go as weather changes. These disappearing and reappearing bodies of water are called vernal pools. During much of the year or for many years in a desert there may not be any water. But when the rains do come, the eggs that are in the sand hatch. Because the eggs are tiny, when they are dry they can be blown by wind to many places. They can also stick to legs of other creatures, like birds and frogs, and go where those creatures go. In this way, Triops eggs are carried to new places, and are found in deserts and vernal pools all over the world.

Triops are also found in many rice fields in the United States, Europe and Japan. In these places, Triops are considered pests because they disturb and eat the roots and leaves of young rice plants and destroy large quantities of crops.

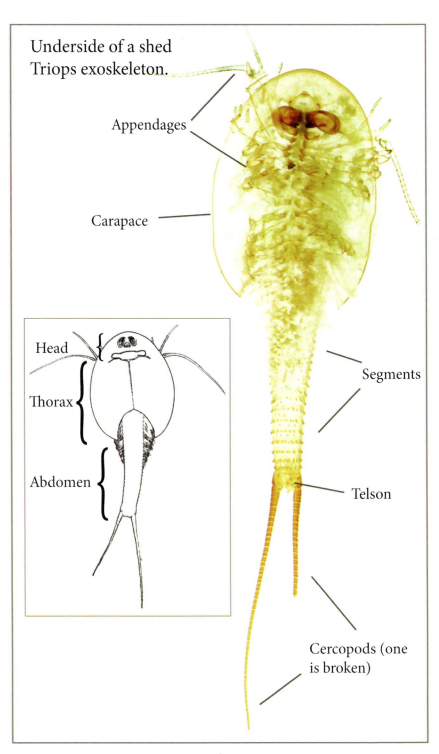

Underside of a shed Triops exoskeleton.

Appendages

Carapace

Head

Thorax

Abdomen

Segments

Telson

Cercopods (one is broken)

What are the body parts of a Triops?

Triops are invertebrate animals, a kind that has no backbone or vertebrae. Instead, Triops have an exoskeleton made of chitin, which is firm—giving it structure and protecting the softer body parts. Because it is rigid and doesn't expand as Triops grow, this exoskeleton has to be shed to make room for a larger body and more parts. This is a characteristic of all creatures with exoskeletons. The process of shedding skin is called molting. With Triops it happens very quickly and the shed skin is almost an exact replica of the creature.

Identifying body parts is critical to studying any form of animal life.

The three main parts of the Triops are head, thorax and abdomen. The carapace or "shield" covers much of the head and thorax.

The thorax and abdomen are made up of segments (similar parts that are repeated over and over) connected by thin membranes, allowing for flexibility in movement. (Some scientists differ on the location of where the thorax ends and the abdomen starts, but all agree that the parts exist!)

The shape of the telson, at the end of the abdomen, is important in telling one kind of Triops from another. The anus, where the waste is ejected from the body, is on the telson.

Appendages or limbs are attached to many of the different segments. They grow out from the body and have different shapes and purposes: long and thin for sensing objects in the water; jointed for walking, swimming and collecting food; flat and thin gills for breathing; and cercopods (segmented flexible tails) for steering. Other names for cercopods are "caudal rami" and "uropods."

Compound eyes

Mandibular groove

Naupliar eye position

Dorsal organ

Cervical groove

Maxillary glands

How do their bodies work? How do they see and eat?

The carapace is the largest part of the Triops and shaped much like a shield. A common name for Triops is "shield shrimp"; another is "tadpole shrimp."

The formal name *Triops* is from the Greek words *tria* and *ops*, meaning "three" and "eyes."

These eyes are near the front of the carapace. The two large eyes are each compound eyes—the kind with many lenses.

Butterflies, grasshoppers, honeybees, dragonflies and most insects have compound eyes. So do many marine creatures, like crayfish, shrimp and crabs.

The third eye in the Triops is different. It is inside the body and called a naupliar eye. On the bottom side of the Triops there is an area like a window (between the two black compound eyes) that lets light into that eye and helps the Triops tell if it is right side up or upside down.

The dorsal organ, slightly behind the compound eyes, may be for chemoreception, or to taste and smell.

The mandibular groove marks a division in head segments, and the cervical groove is the dividing line between the head and the thorax.

Maxillary glands on each side of the carapace control the water pressure inside the Triops' body.

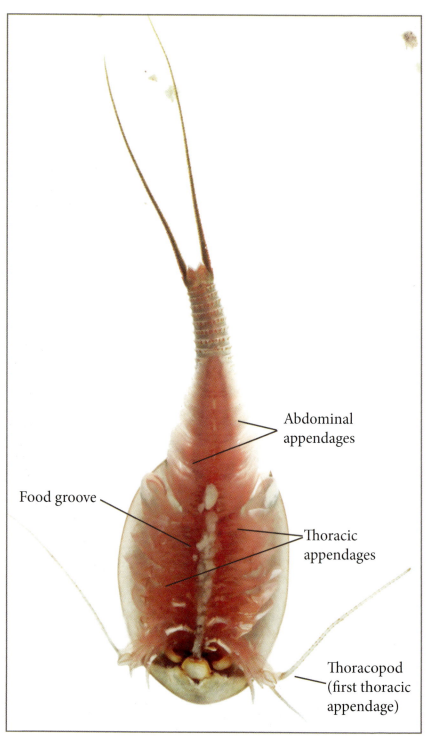

Abdominal appendages

Food groove

Thoracic appendages

Thoracopod (first thoracic appendage)

The underside, or ventral side, of the image on the opposite page shows the Triops with two full cercopods, or tails, and an overview of the thoracic and abdominal appendages. It also shows the food groove, with food being passed to the jaw area where it will be crushed. (Keeping Triops as pets is not dangerous to humans, though. Even if they are taken from the water and held, the jaws are so small they could not hurt us.)

The first thoracic appendages (thoracopods) are different from the rest because of how they split and divide into separate branches. These act like feelers, similar to whiskers on a cat. They can tell if particles they touch are possible food. Triops have antennae, but they are extremely small and visible only under a microscope.

Mandibles (jaws)

What about the other appendages? Are they legs? How do Triops move and breathe?

Other moving parts under the carapace are the appendages used for swimming and collecting food. Like legs, they can move the creature along the bottom of a pool or pond, and on the underside of the water surface.

Generally, Triops do not stay still. They move in circles and somersaults searching for and gathering food. As the water level in the pool, pond or container goes down and as the oxygen in the water decreases, the Triops will walk on the underside of the water surface where the level of oxygen is higher.

Closer to the abdomen are flat, leaf-shaped appendages. These are gills and help the Triops breathe and collect oxygen. In some Triops, hemoglobin in the blood also helps it get oxygen from the water and makes its blood red.

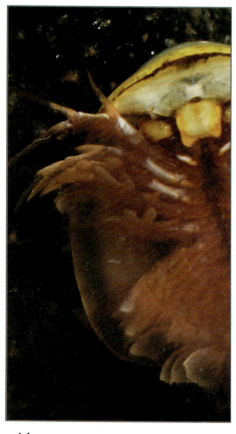

Triops are part of a class of animals called Branchiopods, nicknamed "gill-foot." This formal name is from Greek words related to breathing and feet.

(There is more information about classes in the taxonomy chart later in this book.)

In this photo, taken of the underside of the Triops, you can see the development of the brood pouches or egg sacs on a female. They grow at the position of the eleventh appendages. In this image, like some others in this book, some appendages are blurry because they moved too quickly for the camera to freeze the motion.

How do Triops mate and reproduce?

Some Triops are female, others are male, and some are both
(hermaphrodites). Females have eggs, males can fertilize eggs
and hermaphrodites can have eggs and fertilize them too. When
the female Triops has eggs that don't need fertilization, it is called
parthenogenesis. Not needing separate males and females to
reproduce is common among invertebrates.

Where can I see a real Triops? Do I have to go to a desert during the rainy season?

It would be a fun trip, but Triops can be raised anywhere a small habitat similar to a desert can be maintained for about two months. With a little attention, they can easily be hatched and kept as pets. In fact, they are wonderful creatures to keep as first-time pets: their naturally short life span requires only an equally short commitment from their keeper.

Eggs can be purchased in small packets alone or as part of kits designed for science experiments from companies that specialize in Triops. The basic requirements are simple and egg packets include detailed instructions. Kits are fun too, but really all that is needed is a small container, a desk lamp, daylight, pure spring water and a warm place for your habitat. Carrots are optional but highly recommended for your Triops.

If you add clean sand (like sandbox sand) to the habitat, you can watch the Triops dig and burrow like they would in the desert.

What is the life cycle of Triops?

From the time the eggs hatch to when the Triops die is about three to six weeks.

When the eggs first hatch the larvae are extremely hard to see with human eyes. The best way to see them is to stare at the water and look for little specks of white that are constantly wriggling. But because they are so small, it may not be possible to see them until they are three to four days old. The larvae grow very quickly, doubling in size many, many times in the first two weeks. During this time, they rarely stop moving. They are looking for food and eating. Any small plants or animals—like mosquito larvae, daphnia, fairy shrimp or even other Triops—are possible food. If you are raising Triops at home or school, make food available at all times by keeping a small slice of carrot in the water. In about two to three weeks, the Triops will reach full size.

In a natural environment, Triops are also food for other creatures, like birds that stop at the vernal pools to drink and eat. They are an important part of the food chain. Do not try to raise Triops in an aquarium with other fish because they may get eaten.

Triops larvae, less than 24 hours old (metanauplius larvae). Development of the cercopods can be seen in the image on the right.

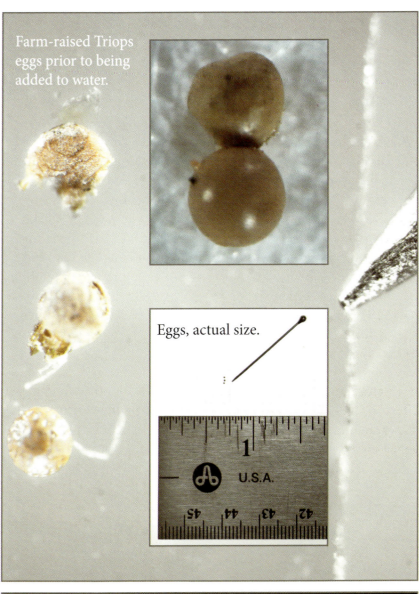

Farm-raised Triops eggs prior to being added to water.

Eggs, actual size.

1
U.S.A.

42 43 44 45

Eggs in the bottom of a tank without sand, prior to drying out.

Will Triops live longer in my habitat than in their natural environment?

They could, but it would be unusual. If one of your Triops has egg sacs, you could try to raise its offspring. If you want to continue raising Triops, save the sand from your container and let it thoroughly dry out. Keep it covered so it doesn't get contaminated or spill. When you are ready to experiment again, add fresh, pure spring water and restart the process.

Should I be sad when my Triops dies?

Every animal has a natural life span—some long and others quite short, like the Triops. No matter how well you look after your Triops, their natural life span will run its course. Try not to feel sad about this. However, once you have watched creatures grow and develop, it is completely normal to miss the excitement of learning about and caring for them.

Even though it may seem strange that most Triops live for just 20–40 days (and sometimes longer if the water in their habitat has not dried up), it may seem equally strange that the species has remained unchanged for over 350 million years.

What makes the Triops such a survivor?

Adaptation—or adjusting to the environment—is the big reason that the Triops species has been unchanged and has lasted so remarkably long.

Examples of Triops adaptations include:

1) eggs that don't die through long dry periods.

2) eggs that hatch in temporary pools because those are environments with fewer predators.

3) being able to reproduce without both males and females and with hermaphrodites. This way, if the vernal pool doesn't have both sexes, the eggs may still hatch.

4) being able to live an entire life cycle during a short period of time means there are greater chances of a next generation.

A Trilobite fossil.

When were Triops first on the earth?
How do we know for sure?

Records of Triops are found in fossils that scientists think are about 300 million years old.

A fossil is the remains of a plant or an animal that has been preserved in rock. It is created when the body of a creature or plant is covered by many layers of sand or mud. With lots of pressure over thousands of years, and when minerals entered the remains, rock formed around whatever was buried in the sand and mud. What was once alive ends up like a drawing or a sculpture in stone.

Trilobites are a large category of creatures that were alive when Triops were first alive, but are now extinct, or no longer alive on this planet. They have been widely studied because there may be more fossil records of Trilobites than of any other creature!

The name *Trilobite* sounds similar to the name *Triops* because they both start with "tri," but there is a big difference. The "tri" or "three" in Triops refers to the number of eyes, while for a Trilobite it refers to the number of lobes or parts of the creature when split down its length.

There are similarities between Triops and Trilobites besides the "tri" prefix. Can you see them?

A nickname for Triops is "living fossil." Can you figure out why?

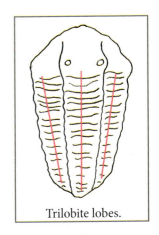

Trilobite lobes.

Dinosaurs were prehistoric creatures too. Did Triops and dinosaurs live together?

Fossil records have been found of Triops both during the Paleozoic Era, long before dinosaurs existed, and during the Mesozoic Era, when dinosaurs were alive.

Why are Triops alive now and dinosaurs are not?

Triops eggs are very special because they dry out and become dormant or inactive. Their development slows down and the eggs enter a period called diapause. In a way the eggs are similar to plant seeds and can be stored for many years, decades even, before water is added and the eggs hatch.

It is unknown for sure what killed all the dinosaurs. But it is thought by scientists that there could have been widespread disease and illness or that the habitat the dinosaurs needed for survival changed drastically. There could have been an asteroid that hit earth or other extreme changes in weather, or both, that killed the food the dinosaurs ate. Sunlight and oxygen could have been restricted causing the dinosaurs to die. Because Triops eggs can become inactive and survive in extreme temperatures (cold and hot) and without light for a long time, they outlasted whatever killed the dinosaurs.

Another name for diapause is suspended animation. In addition to being able to halt development, the eggs can withstand heat to almost boiling (just under 100 degrees Celsius or 212 degrees Fahrenheit). Also, the eggs will not hatch until somehow they know that the water is deep enough to be a good environment for their kind of life.

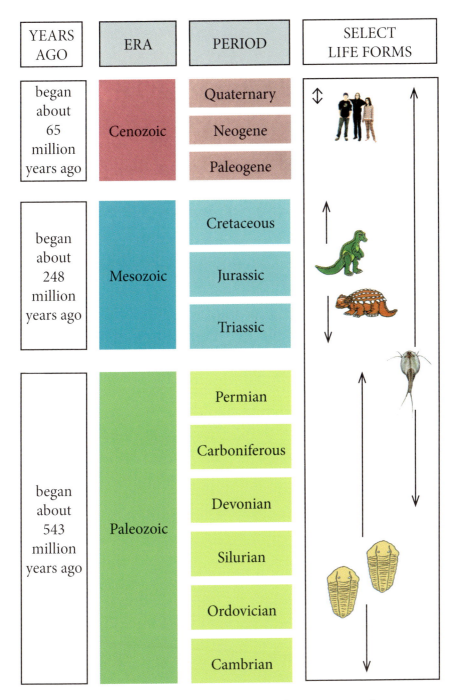

YEARS AGO	ERA	PERIOD	SELECT LIFE FORMS
began about 65 million years ago	Cenozoic	Quaternary Neogene Paleogene	
began about 248 million years ago	Mesozoic	Cretaceous Jurassic Triassic	
began about 543 million years ago	Paleozoic	Permian Carboniferous Devonian Silurian Ordovician Cambrian	

This chart shows Eras and Periods. Other divisions, not shown, are Eons (longer than Eras) and Epochs (shorter than Periods).

Will Triops always be around the way they have been?

It's uncertain. One kind of Triops found in California, *Lepidurus packardi*, is already considered an endangered species. Another kind of Triops, *Triops cancriformis*, is an endangered species in many European countries because of habitat loss.

The future of Triops depends on their environment. Will it stay the way it is? If it does change, will the Triops change too? Deserts are areas of the earth that humans find hard to live in. As a result, they often change the landscape to suit their needs, much as they've changed other environments like river valleys, rainforests and grasslands. If deserts continue to have rainy seasons, though, perhaps the Triops will continue to exist.

What other animals are related to Triops?

Triops are in a subgroup of Branchiopods called Notostraca. Branchiopods are part of a larger group called Crustaceans, and Crustaceans are part of an even larger group called Arthropods. The closest relatives to Triops are fairy shrimp, brine shrimp (sea monkeys) and water fleas.

Hermit crabs, like this stareye crab, are also crustaceans, but are part of a different class, Malacostraca, and within that, they are a kind of Decapod. This type of hermit crab (*Dardanus venosus*) grows to about 3–5 inches (8–13 centimeters), and like other hermit crabs, it adopts and carries around a shell for temporary shelter and protection.

Stareye hermit crab (*Dardanus venosus*).

Caribbean spiny lobster (*Panulirus argus*).

Lobsters are also crustaceans, part of the Malacostraca class and Decapod order. This is a Caribbean spiny lobster (*Panulirus argus*) that grows up to two feet or roughly one-half meter. Like many crustaceans, the eyes of lobsters and hermit crabs are on stalks that grow from the head of the animal.

Yellowline arrow crab (*Stenorhynchus seticornis*).

Other relatives of Triops in the Decapod group include the yellowline arrow crab (*Stenorhynchus seticornis*) and the Pederson cleaner shrimp (*Periclimenes pedersoni*). The yellowline arrow crab is unique because it carries food on the arrow between the eyes. Its body grows to about 1½–2 inches (4–5 centimeters) long. The yellow and blue Pederson cleaner shrimp gets its food off of other creatures, cleaning them at the same time. Its body grows to a length of about 1 inch (2½ centimeters).

Pederson cleaner shrimp (*Periclimenes pedersoni*).

Horseshoe crabs are also called living fossils. Like Triops, they are species that have not evolved much since the Paleozoic Era. Like Triops, horseshoe crabs molt in order to grow. Pictured here is a shed skin of a young horseshoe crab, photographed showing the top and the underside. It is only about 2 inches (5 centimeters) long.

Horseshoe crab shed skin.

The bottom photograph shows another Trilobite fossil, like the one on page 24, approximately 3½ inches (about 9 centimeters) in length.

Neither horseshoe crabs nor trilobites are crustaceans like Triops, crabs, lobsters and shrimp, but there are still many similarities in their shapes, segmented bodies and appendages.

These creatures are also Arthropods, but belong to the subphylum Trilobitomorpha (trilobites) and Chelicerata (horseshoe crabs) instead of the subphylum Crustacea.

Trilobite fossil.

How does anyone keep track of all these creatures? How do we know they are related?

There is a science called taxonomy. Zoological taxonomists are the scientists who name, describe, organize and classify animals into groups. This organization helps us understand similarities and differences between creatures. It depends much on describing the anatomy or body parts and the functions of these parts. In addition, there is an area of science called phylogenetics, concerned with how living things have evolved—how anatomy has changed over time. This also greatly affects how creatures are classified and organized into groups.

Early scientists and natural historians in these areas include Carl Linnaeus, Charles Darwin and Ernst Haeckel.

As newer technologies have been and continue to be developed, the cells, molecules, DNA and genes of living things are being examined in greater and greater detail. As a result, parts of this science of organization and classification are currently debated and there is not one classification system that is considered perfect.

Where do Triops fit in the Animal Kingdom?

The chart that follows shows the connection of Triops to all other life in the Animal Kingdom. It is mind-boggling at times to try to understand how living things are connected. Illustrations and charts are helpful to organize complex ideas.

The basic categories or ranks are: Kingdom, Phylum (and Subphylum), Class, Order, Family, Genus (or its plural, Genera) and Species.

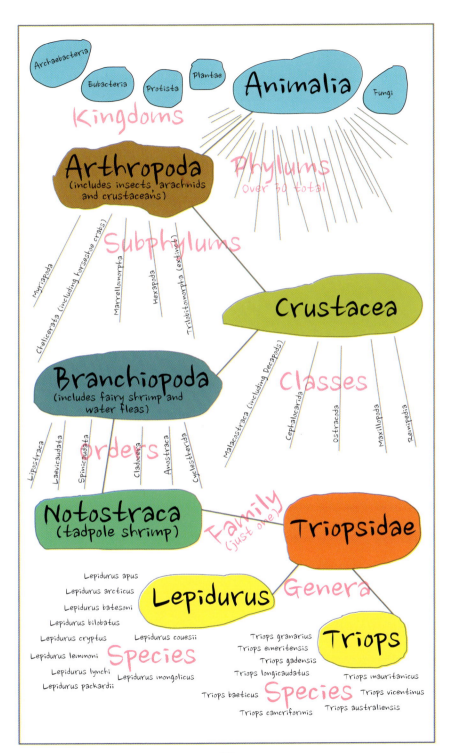

Kingdoms

Archaebacteria Eubacteria Protista Plantae **Animalia** Fungi

Phylums
Over 50 total

Arthropoda
(includes insects, arachnids
and crustaceans)

Subphylums

Myriapoda
Chelicerata (including horseshoe crabs)
Marrellomorpha
Hexapoda
Trilobitomorpha (extinct)

Crustacea

Branchiopoda
(includes fairy shrimp and
water fleas)

Classes

Malacostraca (including Decapods)
Cephalocarida
Ostracoda
Maxillopoda
Remipedia

Orders

Lipostraca
Laevicaudata
Spinicaudata
Cladocera
Anostraca
Cyclestherida

Notostraca
(tadpole shrimp)

Family
(just one)

Triopsidae

Lepidurus

Genera

Triops

Species

Lepidurus apus
Lepidurus arcticus
Lepidurus batesoni
Lepidurus bilobatus
Lepidurus cryptus
Lepidurus couesii
Lepidurus lemmoni
Lepidurus lynchi
Lepidurus mongolicus
Lepidurus packardii

Triops granarius
Triops emeritensis
Triops gadensis
Triops longicaudatus
Triops mauritanicus
Triops baeticus
Triops vicentinus
Triops cancriformis
Triops australiensis

Species

Things to Do

Raise Triops! Keep a notebook with your observations, using words, drawings, measurements, photos and your imagination. Write down your questions and then look for answers.

Make a taxonomy chart! What creatures are you interested in? Can you examine and describe their anatomy? Can you figure out their relatives and why they are connected?

Are there microscopes at your school or library? Can you find out about close-up photography? Can you visit a science museum and observe, ask questions and research the answers? Can you practice sketching so that your drawings share the information you have learned and the beauty you have seen?

Can you look up words in a dictionary or use a dictionary app that will help you understand what you are reading? Can you keep a list of new words or write a story about what you have learned? There are so many things you can do...

Notes and Drawings

Notes and Drawings

Notes and Drawings

Author Notes

The photos of Triops, except the dead Triops in the sand and the egg close-ups, were of Triops underwater. The white background or black background was white paper or black velvet. In general, the clarity of the images is due to using a high-powered studio flash, with a small lens aperture for high depth of field and a high-quality macro (close-up) lens. A tripod was used for many of the images. Images on pages 28–30 were taken by Tony and Marilyn Dahle while on underwater dives in the Caribbean Ocean. The close-up photographs of the dry eggs were taken by Eric Somers at Lori Adams Photo. The dinosaur illustrations on page 27 were drawn by Eloise Adams. All other photos and illustrations are by the author.

The Triops eggs used by the author were from Toyops (www.toyops .com), a resource for farm-raised eggs.

Extensive information about Triops can also be found at MyTriops (www.mytriops.com). Additional reference and inspiration for this book came from *Wonderful Life* by Stephen Jay Gould and other various sources of scientific study. Much practical knowledge and enthusiasm came from Dr. Helen Pashley, zoologist and educator, and Jill Eisenstein, science educator and writer.

* * * * *

More about the author, including other publications and a wide range of photographs, can be found at
www.loriadamsphoto.com.

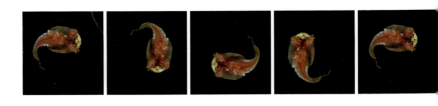